The Sinister Universe:

One force, two fermions, three fields,

four dimensions, and five fundamental constants

Greg Feild

March 1, 2017

About the author:

   I earned a Ph.D in experimental high energy physics from the Pennsylvania State University working on HERA at DESY in Hamburg, Germany studying photoproduction and deep inelastic scattering in electron-proton collisions.

   I did my postdoctoral studies with Yale University working at Fermilab on the CDF experiment at the Tevatron. My primary research interest was particle hadronization in quarkonium production in proton-antiproton collisions.

   Nowadays, I fancy myself a theoretical physicist; although by no means a mathematical physicist nor a phenomenologist! More like a natural philosopher of old perhaps ...

Abstract:

This book is an overview of the 'universal model' of physics, which we have recently derived or induced over the course of five previous books; warts and all!

The universal model aspires to be a fully relativistic, quantum mechanical field description of all particle interactions. It should work on all scales; from the microscopic to the macroscopic to the cosmic. It also seems it should be gauge invariant and renormalizable.

This 'new' model is essentially the current standard model with a lot pruning and a few new assumptions; mainly concerning the mass and magnetic moment of the neutrino, and the role of the relativistic particle mass as an energy dependent gravitational charge.

In the universal model, we assume all calculations can be done in the framework of QED, where the total coupling strength is now due to a combination of the electric charge and the gravitational charge (i.e. relativistic mass) of a particle.

In this current book, in addition to providing a summary and synopsis of the current status of the universal model, we will continue to examine further simplifications that might be made in physics and in our understanding of the world.

physics is fun!

G_F

Apologia:

This book is an attempt to summarize the 'universal model of nature'. This theory has been hashed out in five previous self-published books (^1,2,3,4,5), for a total of, perhaps, one or two new insights into the workings of the world. These previous books also contained several errors which we will try to address here, without introducing any additional ones!

Loyal readers will know that I considered these books a series of papers that I'd dash off as soon as I thought a particular set of ideas had gelled, although they might not have been fully formed.

I wanted to be sure beat my competitors (i.e. readers!) to publication! I jest a bit, but I think the time is ripe for this type of theory and I am sure many others are thinking along similar lines.

Basically, the universal model is a winnowing and grafting together of current models of the world, with the more speculative bits removed, and a few new speculations put in their place!

We will continue, in this book, to critically examine ideas such that space contains gravitational energy and momentum *in addition to* hosting numerous quantum fields, each also carrying energy and momentum.

The universal model is a quantum mechanical theory of gravity and thus replaces the general theory of relativity; so we are no longer 'saddled' with gravitational potential energy stored in space, or the curvature of spacetime.

In a similar way, we will try and examine all the energy assumed to lie in quantum fields, and perhaps, find a way to minimize this feature in a suitable and agreeable fashion.

a feild theory   :)

Introduction:

Space is no longer curved. More specifically, space is not considered to be a substance, or an object, or a collection of interacting objects, or a fluid, or any variation on the electromagnetic ether. Space is empty and void of energy and momentum.

In the 'universal model', the energy of any system of particles, quantum or cosmological, is carried in the mass and kinetic energy of the particles (or planets, etc.) involved. We will continue to speak of potential energy and forces as before, but neither of these are thought to be a property of space (or fields), but rather of the relationships between objects.

Not to worry, we haven't brought back "action at a distance"!

All particles interact by emitting or exchanging photons. These particles carry the energy and momentum. The fields describe the behavior of these particles (or their effects) at a point in space and time. (We will explore this distinction in more detail in a later section.)

Space is the distance between objects as measured in your favorite Cartesian (or other inertial) coordinate system, and time is read off by the ticking of the clock. This is the definition of time and space in quantum field theory (Minkowski space), and the time and space of our everyday lives. In the universal model, Minkowski space is the appropriate representation for all time and space, including outer space. It seems space is Euclidean after all!

The universal model of particle interactions is essentially the standard model stripped to the bone. It works at both quantum mechanical and astronomical scales because the gravitational coupling charge is considered to be a particle's *relativistic* mass. In this model, accelerating objects are allowed, and necessary, as acceleration is a determining factor in the gravitational mass of an object. For example, a spinning galactic object has ('extra') relativistic mass due to its rotation.

Despite having claimed a grand unification of the four forces, we essentially still have two forces; electromagnetism and gravity, and two separate coupling charges; the electric charge and particle mass. However, both of these interactions are expected to be calculable in the framework of QED, and indeed couple to the same field; the electromagnetic field.

Even though electric charge and mass are assumed to couple to the same gauge boson (the photon), we are still left with two coupling constants: alpha and G. This is 'sad', but reflects the one lone asymmetry left in our model; the electric charge.

Let's call this e-Symmetry!

The fundamental particles:

In the universal model there are two fundamental fermions, the electron and the electron neutrino; and one fundamental boson, the photon. From these building blocks, all ordinary matter is formed

The electron neutrino is thought to be the fundamental unit of mass. It is considered to be a 'true' point particle of spin ½. From a consideration of the relative strengths of the electromagnetic and weak interactions, the rest mass of the neutrino has previously been derived (1) to be

$$m\_v = m\_e/e \tag{1}$$

where m_e is the electron mass and e is the value of the electric charge.

The electron neutrino is also believed, from arguments of beauty and symmetry (2), to have a magnetic moment given by

$$mu\_v = e*hbar/(2*m\_v*c) \tag{2}$$

In our model, the motion of mass is considered to be the ultimate cause of all magnetic and electrical forces or fields, hence the introduction of the constant e, even though the neutrino is not electrically charged!

The 'spinning mass' of the (point) neutrino gives rise to its magnetic moment.

In addition, the electron neutrino is thought to 'spin to the left', while the antineutrino spins to the 'right'.

In the universal model, leptons now have two spin degrees of freedom! (?)

All leptons have a 'fixed spin' of left or right, as well as an 'interaction spin' which can point up or down. Particles spin left and anti-particles spin right. Particles maintain their defining direction of spin (left or right), while the spin ½ component can flip up or down during an interaction.

So, in our model the neutrino has an antiparticle. Neutrinos spin to the left and antineutrinos spin to the right. Similarly, electrons spin to the left and positrons spin to the right. This explains why we only observe left handed neutrinos in interactions with the electron.

Matter is left handed. Antimatter is right handed. A sinister universe, indeed!

Since we consider the electron neutrino to be the fundamental unit of mass, we shall now invert equation (1) to obtain a formula for the electron mass in terms of the neutrino mass and the value of the electric charge

$$m\_e = e*m\_v \hspace{5cm} (3)$$

The neutrino mass can be taken from measurement or, ideally, calculated from first principles!

We hypothesize that the electron neutrino is the "self-gravitationally" bound state of 'one quantum of action', h. (Perhaps this mass calculation would be a nice dumping ground for all the infinities of renormalization ....)

Using equation (3), the magnetic moment of the electron can be written in terms of the magnetic moment of the neutrino and the electric charge

$$mu\_e = (1/e)*mu\_v \hspace{4cm} (4)$$

In our new model, the electron is 'not quite' a point particle, although it is considered to be an electric point charge! The (extended) rotating mass of the electron is thought to give rise to the point electric charge in a process we have dubbed "quantum mechanical electromagnetic induction" (3). This electrical energy seems to contribute to the electron rest mass, and the value of e seems to be tied to the quantization of the mass of charged particles as demonstrated by equation (3).

Similar relations are assumed to hold for the muon and the muon neutrino as well as for the tau and the tau neutrino. Hence, we can easily predict the masses of the muon neutrino and the tau neutrino from the measured masses of the muon and the tau.

A measurable prediction of the universal model then, are the *exact* values of the neutrino mass and magnetic moment for all three known particle families, derived from the very simple math of equations (3) and (4).

We can also now understand the *conservation of lepton number* as a generalization of the conservation of electric charge to the concept of *conservation of electromagnetic charge*.

For example, although the electron neutrino does not carry electric charge, its magnetic moment is *huge* (2) compared to that of the electron; the point being that the total electromagnetic potential energy (or total 'electromagnetic coupling capacity') of the electron and the neutrino are *equivalent* (4). This equivalency will only become apparent in high energy particle interactions, and/or in a strong external magnetic field.

This seems to be our broken symmetry or e-Symmetry.

Particle interactions:

Classically, the complete relativistic Lorentz force on a particle due to a specified distribution of charge and mass is now given by

$$F = (me')*E + (me')*vxB - m*F\_g - m*(vxB\_g) \qquad (5)$$

where m is the relativistic mass of the particle, e' is the charge to mass ratio of the particle

$$e' = e/m\_rest \qquad (6)$$

and m_rest is the rest mass of the particle.

The term F_g is the Newtonian force due to gravity and is calculated in the usual way (as tiny as it may be!)

$$F\_g = G*m1/R^2 \qquad (7)$$

where R is the distance between our particle and a gravitational source charge of relativistic mass m1.

The minus sign in equation (5) is to make the gravitational force attractive (regardless of the sign of the electric charge, which determines the direction of the force due to the electromagnetic terms). In previous books (3,4), we had neglected the minus sign in presenting equation (5). This was an error.

The term B_g (very tiny!) is the 'gravitational magnetic field' vector, analogous to the electromagnetic term B, and is calculated in a similar way;

$$B\_g = (G/c^\wedge)*m1*(v1xR)/R^3 \qquad (8)$$

where v1 is the velocity of the source particle. The constant giving the strength of the force, G/c^2, is chosen to give our resulting gravitational waves the speed of light.

Equation (5) for the total force on a particle is manifestly Lorentz invariant due to the common mass factor appearing in all four terms. The electromagnetic terms reduce to the usual ('classical') expression in the low particle velocity limit, where m = m_rest.

The original motivation for 'symmetrizing' the classical Lorentz force with respect to gravity was to prop up a quite separate hypothesis, which was that the neutrino had a magnetic moment proportional to the electric charge e, as shown in equation (2), even though the magnetic moment arises solely from the 'rotating' motion of its mass (2,3).

However, this symmetrization also allows for a classical description of gravitational waves *and* bolsters our conclusion from quite separate arguments (1), concerning a gravitationally bound state of two neutrinos, that the graviton is a massless, spin 1, boson just like the photon. In fact, we were forced to conclude that the graviton and the photon were the same particle for lack of any distinguishing characteristics! (3).

(The 'gravitons' would continue the electromagnetic spectrum just where the radio waves start to fade away ...)

Quantum mechanical particle interactions:

In our new model, the photon facilitates all interactions between particles, just as in quantum electrodynamics, except that in the universal model the photon couples to the particle mass as well as the electric charge.

The photon couples to a particle's 'total coupling charge', tcc, which is defined, rather clumsily (and proffered with a minus sign relative to reference (4)), by

$$tcc = (m/a)*(e' - b) \qquad\qquad (9)$$

where m is the particle's relativistic mass and e' is the charge to mass ratio as before, and

$$a = (4*pi*episilon\_0*hbar*c)^{1/2} \qquad\qquad (10)$$

and

$$b = (4*pi*epsilon\_0*G)^{1/2} \qquad\qquad (11)$$

The tcc is designed to yield

$$alpha = e^2/(4*pi*epsilon\_0*hbar*c) \qquad\qquad (12)$$

for purely electromagnetic (and non-relativistic) interactions, and

$$alpha\_G == m\_e^2*G/(hbar*c) \qquad\qquad (13)$$

for gravitational interactions. (There will be an 'unfortunate' cross term between alpha and alpha_G in charged particle interactions, which we will discuss later on.)

For quantum mechanical calculations, the propagator for the interaction between any two particles is then

$$f(q) = (tcc\_1)*(tcc\_2)/q^2 \qquad (14)$$

where q is the four momentum of the exchanged photon.

We can see from equations (13) and (14), that the energy scale dependence of an interaction is now determined by the mass of the interacting particles, rather than by the mass of the propagator as described in the standard model. In our new model, the propagator is always massless.

Finally, the photon also has a coupling charge, m_g, given by

$$m\_g = hbar*nu/c^2 \qquad (15)$$

where nu is the frequency of the photon and c is the speed of light.

Nonrelativistic quantum mechanical gravitational interactions:

We began our enquiries in reference (1) by considering a gravitationally bound state of two identical electron neutrinos; neutrinium. We assumed our two neutrinos would be bound by the classical Newtonian gravitational potential and thus based our neutrinium model on the analogous, and well known, positronium system bound by the Coulomb potential.

The Bohr radius was derived (missing an all important factor of G, here remedied!) to be

$$R\_neutrinium = 2(4*pi*hbar^2)/(G*m^3) \qquad (16)$$

where m is the neutrino mass. We then compared this result to that expected from general relativity,

$$R = 2*h^2/(G*m) \qquad (17)$$

Noting the discrepancy of a factor of $1/m^2$ between the two results, we had to conclude that the general theory of relativity was incorrect (1,2).

It had a good run!

The strong force:

In the universal model, the strong force is merely the relativistic gravitational interaction balanced by the electromagnetic interaction, plus an additional interaction due to our surprising new 'cross-term'!

Let's look at the total coupling between two electrons. For the sake of simplicity, and for illustrative purposes, we will begin with the non-relativistic case, where m*e' ==> e.

The total coupling, (tcc_e)^2, is then, from equation (9)

$$(tcc\_e)^2 \ = \ (1/a^2)*(e^2 - 2*e*m*b + m^2*b^2) \tag{18}$$

Using equations (10) - (13), we have

$$(tcc\_e)^2 = alpha - 2*e*m*b/a^2 + alpha\_G \tag{19}$$

We shall call the cross term, alpha_strong, where

$$alpha\_strong \ == \ (G/(4*pi*epsilon\_0))^{1/2}*(2*e*m/hbar*c) \tag{20}$$

Remembering that equation (20) is non relativistic, we can replace e with m*e' to obtain

$$alpha\_STRONG \ = \ (G/(4*pi*epsilon\_0))^{1/2}*(2*e'*m^2/hbar*c) \tag{21}$$

We see from equation (19) that our new strong force term acts as a counterbalancing force, changing sign for different combinations of electrons and positrons. (Honestly, I'm a little confused about the role of the plus and minus signs …)

The coupling alpha_strong would be appropriate for describing interactions between nucleons (protons and neutrons), while alpha_STRONG would be the choice for parton interactions within the nucleons.

Having established the existence of this important new coupling term, we will immediately ignore it as we look at possible, very simple models of the proton and the neutron!

The running of alpha:

At relativistic energies, alpha becomes

$$alpha \ = (m*e')^2/(4*pi*epsilon\_0*hbar*c) \tag{22}$$

The proton:

In the universal model, the proton is considered to be the bound state of two positrons and an electron.

As a simple model, we can imagine the system behaving as a simple harmonic oscillator; an idea we first presented, rather badly, in reference (5).

In our revised SHO model, the two positrons would repel one another due to the Coulomb potential, while also attracted to the electron, which would 'sit in the middle'. As the positrons are accelerated by the Coulomb force, their relativistic masses will increase, leading to the inevitable restorative power of the gravitational force. Of course, the electron will be relativistic as well.

We can see that this model allows for both asymptotic freedom and confinement! This is due to the energy dependence of the gravitational coupling charge.

In our strong force model, the confining potential is due to Newtonian gravity and an energy dependent particle mass rather than an additional term linear in the separation R.

The equilibrium, or 'ground state', of the proton would be where the Coulomb force balances the gravitational force, or

$$e^2/4*pi*epsilon\_0 \ = \ Gm^2 \qquad\qquad (23)$$

We can now solve for the relativistic mass of the electron and positrons inside the proton

$$m \ \sim= \ e/(4*pi*epsilon\_0*G)^{1/2} \qquad\qquad (24)$$

The neutron:

Our neutron is composed of an electron, a positron, and an electron antineutrino; so a simple model is not quite as easy to conceive!

However, we suggested in reference (5), that the electron antineutrino may 'orbit' a bound positronium nucleus. This led to the idea of the inter-nucleon bonding force being due to the sharing of neutrinos, or the overlapping of 'neutrino clouds'; a.k.a the ionic model!

This seems a good time to introduce alpha_weak as a function of the neutrino mass

$$alpha\_weak \ == \ m\_v^2*G/hbar*c \qquad\qquad (25)$$

The weak force:

In the universal model, weak interactions always involve the neutrino, and thus are weak due to the tiny neutrino mass.

This view of weak interactions has been presented in some detail in reference (4); although I have since come to realize that the muon decay model presented there was "not even wrong"!

Here, we will summarize our new model of the weak interaction with a look at the classic example of neutron beta decay, and then propose a more reasonable (or at least less wrong ...) Idea for the process of muon decay.

In beta decay, the neutron turns into a proton

$$n \; ---> \; p + e- + v\_e^bar \tag{26}$$

In the standard model, beta decay occurs when a down quark emits a massive W- boson to become an up quark. The W- then decays into an electron and an electron neutrino

$$(u,d,d) \rightarrow (u,u,d) + e- + v\_e^bar \tag{27}$$

This process is highly suppressed due to the mass of the W and not due to the nature of the weak charge.

In our new model, where partons are leptons, beta decay now looks like this

$$(e+,e-,v\_e^bar) \rightarrow (e+,e-,e+) + e- + v\_e^bar \tag{28}$$

Here, the antineutrino emits a photon which then decays into an electron and a positron. The coupling at the antineutrino vertex is alpha_weak, which suppresses the interaction.

For muon decay, we imagine the muon emitting a muon neutrino just as in the standard model, but rather than coupling to a virtual W, the muon couples to a virtual electron, which emits an electron antineutrino, thus becoming a real electron.

The muon (and, of course, the tau) is the only 'non-composite' particle that decays!

Is the muon a massively excited electron? An electrically excited muon neutrino? Both? Are there two paths to the muon?

The electroweak bosons:

The massive bosons of the standard electroweak theory are no longer thought to be 'field quanta', but are now considered to be resonant bound states of the two fundamental leptons.

W+ = (e+, nu_e)

W- = (e-, nu_e^bar)

Z   = (e+, e-)

The Z looks to be an (higher) excited state of positronium; the first excited state would be assigned to the pion. Or perhaps, the Z could be made up muons instead, or an admixture of all three (known) charged leptons!

In this model, the newly observed Higgs' boson would then be a bound state of an electron neutrino and an electron antineutrino.

H   = (nu_e, nu_e^bar)

The vacuum:

In the universal model, the Higgs boson is no longer required to provide the particles with mass, and indeed, the universe is no longer permeated by the Higgs field (or the W field, or the Z field, or the quark and gluon fields, etc.!).

In the standard model, the Higgs boson was thought to arise from a symmetry breaking of the energy of the vacuum.   Now, this apparent feature of the 'physical vacuum' no longer seems to be necessary.

In our new model, the vacuum has no energy and no symmetry (or asymmetry for that matter!) and is not a teeming cauldron of subatomic particles.

In fact, we postulate that particle-antiparticle pairs cannot appear out of the vacuum in the absence of matter. Of course, this is a moot point since we would require matter in order to detect the particle-antiparticle pair!

It's a bit of a head scratcher.

The conservation of electromagnetic charge:

Let's now revisit our new idea of the conservation of 'electromagnetic charge'. We shall begin by considering the conservation of angular momentum in particle interactions.

We know that the intrinsic spin angular momentum, ($\frac{1}{2}$*hbar), of the lepton is conserved in all particle interactions. We also know that the spin angular momentum is independent of mass! That is, the intrinsic spin angular momentum is the same for the neutrino, as for the electron, as for the muon, etc.

We would like to construct a mass dependent 'angular momentum' type object that may also be conserved in particle interactions; and so take the product of the only two likely quantities at our disposal; the mass and magnetic moment of a particle. Thus, we define

L_mass == m*mu = e*hbar/2*c   (29)

We see the mass dependence magically disappears, yielding the conservation of electric charge, even for neutrinos!

As the magnetic moment of a particle is a vector quantity, our new conservation law not only demands the conservation of 'electric charge', but also the conservation of the direction of spin.

Another interpretation, is that our new conservation of electromagnetic charge, or lepton conservation, is equivalent to the idea of the conserved electroweak isospin currents in the standard model.

Thus, 'weak isospin' from the standard model is a good and exact symmetry in our new universal model.

So, the electron and the neutrino are similar particles in 'isospin space', and transform into one another under rotations in said space. I'm not sure exactly what this means, physically, but it seems exciting!; perhaps that they differ only by electric charge, or … e-Symmetry.

We can also now explain the conservation of baryon number since baryons are composed of leptons.

The Pauli exclusion principle:

Now that our most fundamental lepton, the electron neutrino, has a magnetic moment, we understand (as most people have long suspected) that the Pauli exclusion principle is due solely to the electromagnetic interaction between leptons -- which act like tiny refrigerator magnets!

Quantum field theory:

So, in the universal model, we have three fundamental particles/fields; the electron field, the neutrino field, and the photon/electromagnetic field(s). The complete Lagrangian for any system of particles would then consist of a Dirac term for the leptons (or maybe two separate terms), a Klein-Gordon term for the photon, and the Electromagnetic term for the interaction! That's it!

This may technically be four fields, but I think it's close enough for a catchy book title!

How much energy and momentum should we expect to lie in these quantum fields?

It does not make sense (or, at least, it is not operationally useful) to talk about the force field or potential field of a single particle; its strength, how far it extends in space, etc.; without the presence of at least one other particle somewhere in the universe! It follows then, that a particle need not, and most likely does not, create a force field somewhere where some other particle is not, and in fact may never be.

By this argument, the fields of QFT should be construed or constrained to describe the relationships between particular particles, and thus need not permanently permeate all space and time. One is reminded of the problematical 'test charge' of classical physics …

The $1/R^2$ law:

All matter interactions are due to real photon emission or virtual photon exchange.

Both of these manifestations of the electromagnetic interaction follow the $1/R^2$ law of diminishing returns.

The mechanism for the $1/R^2$ behavior is completely different for the two cases, although the reason for the behavior is exactly the same! Both are dependent on space being flat and three dimensional.

For real photons, the intensity falls off as $1/R^2$ because the net flux of real particles (or energy and momentum) through a sphere of any radius surrounding the source, must be conserved.

For virtual photons, the strength of an interaction between any two particles diminishes as $1/R^2$ as a consequence of the uncertainty principle.

In either case, space must be Euclidean.

Photoproduction:

My graduate thesis was a study of the 'photoproduction' of hadronic jets in electron-proton collisions.

In photoproduction, one studies the murky region where the 'virtual' photon exchanged between an electron and a proton in a collision is 'almost on-shell', and thus the scattering can be analyzed as a collision between a proton and a real photon! Our measurements concerning these events (cross sections, etc.) agreed, on the whole, quite nicely with the theoretical predictions.

Besides the prophetic irony as regards our current thesis, I mention this subject now to illustrate that although we usually speak of real or virtual photons as facilitating an interaction, in many cases this will be a fine distinction, and it may be more natural to imagine a continuum from virtual photons to real ones.

Cosmology:

The general theory of relativity is no longer thought to describe the gravitational interaction of matter. Hence, many of the current problems in cosmology, which imply the inadequacy of the current theory, need to be reevaluated in the light of our new theory.

Large scale neutral matter interactions are described by the cosmological Lorentz force

$$F = m*F\_g + m*(vxB\_g) \tag{30}$$

It is important to note, that the mass of the objects involved is the *total relativistic mass*, which includes the mass due to rotation.

Angular motion is acceleration. It is also translationally invariant and thus considered to be absolute motion. Thus when considering the interaction between two galaxies, for example, one must include the relativistic masses of the individual spinning bodies, as well as the contribution to the total relativistic mass of the galaxy due to the bodies rotating about the center of the galaxy!

A body on the edge of a rotating galaxy will have a relativistic mass relative to the center, modifying the central force on the body and its angular velocity. No more dark matter!

We can also now see why planets bulge. Matter on the equator of a planet is more massive than matter at the poles.

Red shift:

The gravitational contribution to the redshift of light from a distant galaxy is no longer attributed to the stretching of space. Instead, the shift in frequency is due to the gravitational work done on the photon by the mass of the emitting galaxy. This view will yield new relative velocities for the various galaxies, as well as, perhaps, new mass estimates.

This idea can also account for the cosmic background radiation.

In an eternally cyclic, expanding and contracting universe, matter would change course, but free photons would not. The frequencies of all extant photons would be stretched; first this way, then that!

Black holes:

In our model, black holes consist of photons and electron antineutrinos (3), as well as whatever other detritus happens to get sucked in.

A black hole can be considered to be an 'infinite' spherical potential well of radius R. We imagine that the gravitational potential is constant inside the black hole

$$V(r) = G*M^2/R ; \qquad ; r < R \qquad\qquad (31)$$

and falls off in the usual way for r > R,

$$V(r) = G*M^2/(R+r) \qquad ; r > R \qquad\qquad (32)$$

The energy levels of the antineutrinos inside can be calculated using well known techniques. Since the potential energy of the well is not actually infinite, there will be the possibility for the high energy neutrinos to tunnel their way out!

For the photons, I don't like the idea of 'plane wave' solutions, and so propose instead, a collection of di-photon bound states.

It's all quite speculative at the moment!

The eternal universe:

The big bang is a prediction of the general theory of relativity. Now that we no longer embrace this particular theory, do we still need the big bang?

Would you rather an eternal universe or one created from nothing?

Gravity is not attractive at all scales. As particle separations approach zero, gravity is ultimately repulsive due to the electromagnetic interaction between the individual magnetic moments of the particles.

The big bang, then, is actually only the latest bounce; and is ultimately due to the Pauli exclusion principle!

The five or so fundamental constants of physics:

For those who like their fundamental constants dimensionless, we offer up; alpha, alpha_G, alpha_weak, and alpha_strong. As the fifth constant, we suggest the Lande g-factor for the magnetic moment of the electron; which, of course, we have not discussed. It may seem a weird constant, but it is dimensionless and already measured and calculated!

We also have not yet discussed temperature. But, we will throw in Boltzmann's constant for good measure!

Conclusion:

Space is no longer curved! Such a statement may sound naive, even *counterintuitive*, in this day and age where physics is expected to boggle the mind rather than explain the world.

In our new, more modest model, the universe is not imagined to be a computer, nor a hologram, nor even information; but rather a system of massive particles interacting under mutual forces of repulsion and attraction!

Imagine that!

And your brain just exploded.

Confessions:

I think this is a pretty good model.

It's beautiful, symmetric, and simple; yet not simplistic. It posits two fundamental particles, and explains the origins of their mass and charge, as well as their myriad ways of interacting through a single field; manifesting as apparently four dissimilar forces!

References:

1. "A quantum mechanical theory of gravitational interactions", Greg Feild, CreateSpace Independent Publishing, 8/29/2016

2. "Observations on the quantum mechanical nature of gravity", Greg Feild, CreateSpace Independent Publishing, 10/8/2016

3. "On gravitation and electric charge", Greg Feild, CreateSpace Independent Publishing, 11/1/2016

4. "On spin, mass, and charge", Greg Feild, CreateSpace Independent Publishing, 11/29/2016

5. "On angular momentum, acceleration, and absolute motion", Greg Feild, CreateSpace Independent Publishing, 1/4/2017

Bookshelf:

Elementary Modern Physics
Richard T. Weidner / Robert L. Sells

Introduction to High Energy Physics
Donald H. Perkins

Modern Elementary Particle Physics: The Fundamental Particles and Forces?
Gordon Kane

Gauge Theories of the Strong, Weak, and Electromagnetic Interactions
Chris Quigg

Time, Space, and Things
B. K. Ridley

Space and time in the modern universe
P. C .W. Davies

The Paradox of Cause and Other Essays
John William Miller

Landmark Experiments in Twentieth Century Physics
George L. Trigg

50 things you need to know: universe
Joanne Baker

On the web:

snarXiv vs. arXiv

A Capella Science - Bohemian Gravity!

www.ingramcontent.com/pod-product-compliance
Lightning Source LLC
Chambersburg PA
CBHW051830170526
45167CB00005B/2224